BEE HUNTING

BEE HUNTING

A BOOK OF VALUABLE INFORMATION
FOR BEE HUNTERS—TELL HOW TO
LINE BEES TO TREES, ETC.

BY
JOHN R. LOCKARD

Published by
A. R. HARDING, Publisher
Columbus, Ohio

ISBN 0-936622-00-8

By A. R. HARDING PUB. CO.

CONTENTS

SOME MEMORIES OF BEE HUNTING

I WAS born in a little valley, hemmed in by mountains running north and south on either side. It varies in width from one to three miles from the foot of one range to the other. From my home I have a clear view of these beautiful mountains and, as these mountains and lowlands teemed with game of all kind, and being heavily timbered, made an ideal location for the home of the wild bee. From early youth I loved to lure the wild turkey, stalk the deer and line the bee to his home. Is it any wonder that after forty years of undiminished passion for sports of this kind that I can truthfully say there is scarcely a square rod of these mountains that is not indelibly impressed on my mind in connection with some of the above mentioned sports or pastimes? I will confine myself in this work to the subject of Bee Hunting, believing it to be one of the most fascinating and beneficial of pastimes.

PREFACE

IN THE preparation of this work, it has been my aim to instruct the beginner in the art of bee hunting, rather than offer suggestions to those who have served an apprenticeship at the fascinating pastime. I do not wish to leave the impression that I think others who have made this a study do not know enough on the subject to give suggestions; far from it. But to be candid with each other, as lovers of nature and her ways should be, even though we be veterans in the business, by an exchange of ideas we can always learn something new and of value. Many books on sports of various kinds have been written, but outside of an occasional article in periodicals devoted to bee literature, but little has been written on the subject of bee hunting. Therefore, I have tried, in this volume, Bee Hunting for Pleasure and Profit, to give a work in compact form, the product of what I have learned along this line during the forty years in nature's school room.

Brother, if in reading these pages you find something that will be of value to you, something that will inculcate a desire for manly pastime and make your life brighter, then my aim will have been reached.

I am very truly yours,

JOHN B LOCKARD.

BEE HUNTING

CHAPTER I.

AN OLD BEE HUNTER.

The bee hunters in my early days used one of two methods in hunting the bee. The hunter would select a clear day, generally during buckwheat bloom, and after determining on a course, sun them to the tree. This was done by placing the hat or hand between the eye and sun as close to the light as the eye would permit. If the hunter knew the difference between the flight of a loaded bee and an unloaded one he would keep on the course until the tree was located.

This method must undoubtedly be injurious to the eyes and I do not follow this plan nor advise others to do so. The other method was what was termed burning or baiting. A fire was built near where the bee tree was supposed to be, large flat sand stones were placed on the fire and heated. One of these was removed to some place clear of trees and underbrush, some bee-comb, dampened with water, was then placed on the stone, and when the fumes of the comb would go off into the air any bees flying near were apt to be enticed to the bait, which was sprinkled on a bunch of bushes and laid near the stone. Many bees were found in this way, but if they went any great distance two or more fires had to be built. This would require much time and often the hunter, not being careful in extinguishing the fire, the surrounding leaves would catch fire and a destructive forest fire would result. Therefore it shall be my aim to eliminate anything of an injurious or objectionable nature in the work I lay before the reader.

On a calm morning in the early part of November, I went to the top of the mountain west of my home. The day was an ideal one. The trees had shed their leaves, making a thick carpet over the earth. It seemed that all nature was getting ready for a long winter sleep. All flowers except a few bunches of mountain goldenrod were dead. The bees seemed to be aware that their labors were about ended and were eagerly looking for anything in shape of sweets that would add to their store of supplies and thus help to tide over the long winter. After arriving at the top of the mountain I built a fire, heated a large flat stone and took some bee comb and proceeded to follow the example before mentioned. After watching quite a long time and not seeing any bees I was on the point of giving it up, at this place at least, when that sound so delightful to the ear of the bee hunter, the silvery tone of the bee in flight, came to my ear. Several times the sound was repeated but so far I had not got a sight of it. On looking over the top of the bushes I saw two bees flying slowly, sometimes coming near the bait, then darting away, then returning and finally settling down on the bait. All was anxiety! I must be sure to see these two bees take their homeward flight. In a very short time one of them slowly raised from the bait, circled a time or two, and then darted away so quickly that I knew not where. Now the other one won't escape me so easily. But when I turned to look, she, too, was gone. In a short time they were back and lots of others close behind. In a half hour there must have been a quart of bees on the bait. By this time I had seen a number of bees fly due west and some due east. So taking another hot stone and going some distance on the course west, I put the stone down, burnt more comb, and in a few minutes had lots of bees. They still continued westward. The next time I stopped where a swamp extended from the top of the mountain back some two hundred yards. There were many large gum trees growing in this swamp. After a while I was convinced that the bees flew at right angles from the former course. Leaving the bait I went into the swamp and found them going into a large gum tree about twenty feet from the ground. My spirits were high, this being the first bee I had ever found entirely by myself. Taking out my knife and going up to the tree to put my initials thereon, my spirits fell as suddenly as they had risen. There in plain view were the letters I. W. The spirit of selfishness then showed itself. What right had anyone to take this bee from me? I had almost come to the point of thinking I had a monopoly in the bee hunting business and that others had no right to intrude. I trust others do not show this spirit and am sure I have got rid of it myself. If there is any pleasure or benefit to be derived from anything, God certainly intends it for all. The initials would not correspond with the name of anyone I knew, but supposed that some time I would find out who I. W. was. Now the bee that flew east could be looked for, but what was the use? Hadn't the best bee hunters in the country tried to find it and failed? Beyond a certain point all trees disappeared. This was the only Italian bee known to be in a radius of ten miles and it was not a great while after their introduction into this country. So taking my way to the top of the mountain near the edge of the swamp, I was surprised to find a cabin, and from indication it had just been built. On going up to the door my eye fell on the occupant, a man well up in years. In one corner was a number of steel traps. In another a rifle of the then modern type. These signs told me that a new hunter had taken up his abode among us. He told me to be seated and moved over on the rude bench to make room for me. He began by asking me what I was doing out on the mountain, and as I was so young, no doubt had an idea that I was lost.

I told him that I was bee hunting and had found one but some one had found it before I had, and that the initials I. W. were cut on the tree. Turning to me he said, "You don't know who that stands for? Well, young man, I kin tell you. I. W. stands for Ike Ward, and that's me. The little fellers come sippin' around my cabin and I give 'em a little sweet water and found 'em in a jiffy." I then told him of the Italian bee. He asked me why I didn't find it. The reply was that the very best bee hunters in the country had tried it and failed and I supposed it would be of no use for me to try it. "Well, they must be great bee hunters; why, young man, I would rather undertake to find a bee than ketch a rabbit in a good trackin' snow. The rabbit might jump up and run away, but after I get my bee started, he's mine." It was getting well along in the afternoon and I told him I must go home. "Well, your folks might think something has happened to you and I won't ask ye to stay any longer; but come up again and we will find that yaller bee." I thanked him and asked when it would suit him to go. "You kin come any time you keer to, but ye'd better come early when you do come, fer I might be out scoutin' round and not be home." That proposed bee hunt was the only thing thought of on my way home, the only thought that went with me to my bed, and in my dreams I saw the most beautiful yellow bees in the world on combs of snowy whiteness, some of them as large as a door.

Early the next morning, before the sun had shown himself to the people down in the valley, I was far on my way up the mountain on my way to the hunter's cabin. Great drops of sweat were standing all over my face, but I never slackened my pace until I heard the cheering "Good morning" from the old hunter at the cabin. "Jist come and rest yerself. It's a little too early fer bees to fly yit." I replied that I wasn't tired. "When I was your age I didn't get tired either, but if you get to be as old as me you won't walk so fast up hill; you're all a lather of sweat."

About an hour later we went out to where I had first baited the bees. I began to gather wood to start a fire and burn for them again. "What are ye goin' to do with that wood?" was his inquiry. On being informed that this was the way I got them to bait, he chuckled to himself and said he would show me a better and easier way. He then took a handkerchief from his pocket, then a small bottle containing something that was of a fluid form, and sprinkled the handkerchief with it. He then got a pole eight or ten feet long and put the cloth on one end, raised it as high in the air as he could, moving it back and forth in the breeze. Very soon hundred of bees were darting through the air. The pole was slowly lowered until the handkerchief rested on the ground, sweetened water was sprinkled on some bushes, and in a few minutes the yellow bees were flying east and the black ones found previously flying west.

This was a very simple, but a new departure from the mode followed in those days. He explained to me that the little vial contained water, with a few drops of the oil of anisseed added, and there were other scents perhaps better, but this being the only kind he had at that time was the reason for using it. We went directly east on the course four or five hundred yards. This brought us to the top of the mountain and to a large rock that was fully one hundred feet from the ground at the base to the top. From this rock we had a clear view of the valley below. The eastern side of the mountain was very hilly, and covered with a dense growth of trees, and farther down, this forest never hearing the sound of the woodman's ax, became so dense that the sun could scarcely find an opening to the earth. The cloth was sprinkled with more of the scent, waved a few times in the air, and laid beside the bait. which was composed of sugar and water, on the rock. Bees came in abundance. Very soon we could see some bees, heavily loaded, circle around and dart off down, down, until lost to our sight. Others would fly both north and south along the top, making three distinct courses. The old hunter watched these different flights for a considerable time, then going some distance along the top, and after a short time came back saying, "Just as I expected. These fly out there, make a turn, and come back to join the course that flies straight down. Now come with me out the other way and we will see if the others don't do the same." Sure enough! Taking our station some fifty yards from the bait we could see them coming heavily loaded, bend down and back toward the main course.

"I have found many bees in my time, young man, an' never saw one act this way unless the tree was close. They act like they don't want to leave that rock; but we will go down and look at some of that timber." As all the timber far below had been looked at many times in the past I thought it useless but did not say so. After looking at the nearest trees below, those farther down were examined. The morning had been cloudy but now the sun was bright and clear. The hunter placed his hand before his eyes and gazing up at the sun said he "never saw sich actin'; they seem to come right toward the ground. I have found 'em in queer places but never in the ground." Just then a bee lit on some leaves in front of me. I called his attention to it. "Now ain't it a beauty? Poor little fellow; got too heavy a load an' has to rest. Now watch sharp; when he goes he will likely fly straight." In a short time he slowly raised, made a half circle, darted down the mountain, and was lost to me. Not so with my companion. Stooped low, his arm thrust forward as though guiding the bee in its flight, he slowly turned his arm, still following, until he was pointing straight up the hill. "As sure as my name is Ike Ward that bee flew up the hill, and just as sure its home is there, too."

Up the hill he went, looking more carefully at every tree, until the last tree below the rock had been reached. I was on the upper side of this tree and was almost sure that it must be in this one. The old hunter was on the lower side, gazing intently up the hill toward the rock. For some time he stood thus, then said, "You had better look behind you if you want to find the yaller bee." On turning round I saw a steady stream of bees going in and coming out from the very base of the rock. The mystery was a mystery no longer. They had baffled all the bee hunters in the community for three years, but at last they gave up the secret of their hidden home to Ike Ward.

Taking a piece of paper and writing thereon these words: "This bee was found by Ike Ward and pard; if any person find it please don't mislest it." He laid the paper above the entrance of the bees, and, laying a stone on it to keep it in place, we ended this our first bee-hunt together. This was only one of the many delightful trips which I took with the hunter, only one of the many valuable lessons received from him on this fascinating pastime. He has long since passed away, but the book of nature was open to him at all times and with a spirit that had no taint of selfishness in it, was always ready to impart knowledge to others.

CHAPTER II.

EARLY SPRING HUNTING.

Bees are very fond of salt in the early spring, and, in fact, in all parts of the season when brood rearing is in progress. Now we will start out some fine spring morning, take a hatchet or an ax and a polk of salt, and we will go up on the side of the mountain and chop out a little trough large enough to hold a quart or more, then sprinkle a little water, scented with oil of anise or bergamont, on the outside of this trough, then put a few corncobs and a handful of salt in the trough and place the trough in the fork of a small tree out of the way of any stock that may be pasturing in the woods. Our work is now done at this place. We can go on and put out several of these baits along the mountain. The first rain that comes will fill the trough, dissolve the salt, which will soak into the corncobs, and the scent which we placed on the outside of the trough will entice any bees that may be flying. After this we go home and a day or so after the first good rain that comes, we will go back and the chances are that we will have several good courses. Now we will cover the trough over with a bunch of leaves—green boughs—and sprinkle these freely with sweetened water. Take a pint bottle, fill it one-fourth full of granulated sugar and fill up with water. This is better than more sugar, for when the syrup is too thick it requires more time for the bees to load up and if too thick, in a short time the bushes become sticky.

After several bees have loaded up and gone home, we will take a cloth and saturate it with the same scent used on the trough, then take the bait —bunch of bushes—with us on the course, hunt a place as free from timber as possible and lay out bait on the top of a bush, the cloth beside it, and in a short time we should have plenty of bees. After determining on the course the same tactics are pursued until we arrive at the tree, or, if we have good reason to believe the bee stands in any certain group of trees and we fail to find the tree, to make sure that our ideas are correct we will move our bait off to one side of the original course and thus get a cross course, and at the junction of the first line of flight and this second line, the bees must certainly have their home. We must look at every tree with the utmost care, for it is a very easy matter to overlook a bee tree, even experienced bee hunters have done this. But if we take time to examine a tree from all sides we should always be able to locate them.

CHAPTER III.

BEES WATERING. HOW TO FIND THEM.

As soon as the bees begin to stir in the spring they go searching around for water, for this is one essential element in brood-rearing. Early in the season the ground is generally so full of water that bees are not confined to any certain place in order to get the amount needed. But later in the season,when the ground has dried off and wet weather springs have dried up, if we go into the woods along the mountain and visit the never-failing springs sure to be found in the hollows and low flat places, we will be pretty sure to find bees at some of these places.

It is not often that bees are numerous enough at these springs to make what would be termed a strong course, but by following the plan which I here give, you can, in a short space of time, have all the bees necessary, with no danger of having bees from other trees or from our neighbors' stands, which would make a mix-up, and make it much harder for us to follow the bee that is watering. When we go on a trip of this kind first we will provide ourselves with a small glass tumbler; a cover, made of some dark heavy material, long enough so that when slipped over the glass it will come within one-fourth of an inch of the open end. Then we will take a few drops of honey in a small vial, the scent, cloth, and bait of sugar and water mentioned previously. When we find the bees watering we take the glass, without cover, and place it over the bee, which will immediately try to fly and finding himself a prisoner, will crawl around the upper part of the glass. Previous to this a few drops of the honey were placed on a piece of cardboard or large leaf. Then we lift the glass and place the hand under to prevent the bee escaping and place it on the cardboard or leaf. Now place the black hood over it and watch the result. There is but one place for light to enter and this is the narrow opening at lower end of cover. In a moment the bee can be seen crawling around the bottom, sometimes reaching down to the cardboard. Now he has found a drop of the honey and seemingly forgets his sad plight of a moment ago and proceeds to take a meal. The glass is lifted gently off, the dark thick cover preventing him from seeing our hand. As soon as he is loaded he starts and circles many times and then goes home, and in some manner that we can't explain, tells others of what delicious sweets he has found. No more water for that bee; he is bound to come back and search for more honey.

We can go and catch as many bees as we think it necessary, but generally five or six would be ample. Then the scented cloth is placed on the ground, a bunch of green bushes laid on the spot where the cardboard had been sprinkled freely with sweetened water, and we are soon ready to start on the course, following the instructions given in previous chapter.

CHAPTER IV.

HUNTING BEES FROM SUMAC.

Sumac begins to bloom about the first of July and continues through the month. It is unquestionably the greatest source of honey in the country in which I live. From the time the dew is off until dusk the bee is busy on it. Every old worn-out field is plentifully supplied with it and a different variety is found growing in small patches all over the mountains. I have found more bee trees by the plan now given than perhaps any other.

We will visit some of these places and select a spot where there are a few bunches near together, if no more than a half dozen bunches the better. Now having our bottle containing bait prepared, let us select two or three bunches standing close together and sprinkle them freely with the bait, then break off all others standing near. At first the bees will fly around as if they don't like to light on the wet bushes but the ones that were used to getting honey from these flowers may visit other flowers and fly away, but they are sure to come back, and, after taking a sip, finding it a quicker method of getting a load of sweets, settle down to business and in a short space of time adapt themselves to the new order of things and are soon on their way home, never failing to return. bringing others along. Keeping the bushes well supplied with bait, we will soon discover a course and perhaps two or more. Then take the scented cloth, lay it near the bait, and after ten or fifteen minutes break these bushes off a foot or more below the flowers and we are ready to start on the course. After going two or three hundred yards, select a place clear of trees so that they can fly on their course without being compelled to fly around timber, lay the scent cloth near by, and in five or ten minutes you will have plenty of bees, or, we may be going on the line of flight and find the bees suddenly cease to come to bait. This is an unfailing sign that we have passed the tree or are very close to it.

CHAPTER V.

HUNTING BEES FROM BUCKWHEAT.

During buckwheat bloom, which occurs in the month of August and early part of September many bees are found. Some hunters line them to the tree by sunning. This method requires a very clear day and unless the hunter thoroughly understands this art, knows an unloaded bee from a loaded one, he is not apt to be very successful. Besides this fact I have known many hunters to so injure their eyesight as to become, in old age, partially blind and perhaps altogether so. I, myself, have found many bees in this way and feel certain that my eyesight has been injured, but am very thankful that I discarded this method many years ago.

Bees do their work on buckwheat from the time the dew is leaving until near noon; and on a hot, clear day but few bees, if any, will be found working on it after 12 M. One of the greatest elements of success in hunting bees by the baiting method is to use a scent that is the same as the flower the bee is working on. Therefore, gather some of the flowers of the buckwheat and have them distilled, or, if this is out of the question, put some of the flowers in a quart jar, say half full, well packed down, then just cover with diluted alcohol and let it stand a few days and you have an ideal scent to use at this particular time. After getting a course from a field of buckwheat, about ten or half-past ten go on the course, and when you come to a place clear of underbrush and no large trees to bother the flight of bees, sprinkle some of the scent mentioned above on some leaves and near the scent place a bunch of bushes sprinkled with bait made by filling a pint bottle one-fourth full of honey, one-fourth of granulated sugar and one-half water. Many bees, at this time of day, are going to and fro from the field. Some of them find nectar harder to get than it was an hour before and some fly on the homeward journey lightly loaded. They are beginning to lose faith in the buckwheat field and these are the very ones that detect the scent first. Others are becoming dissatisfied as these first ones did—one rubs against another, and in bee language tells that he has found something mighty good down in the bushes, and by the time the bait is licked up we should have a direct course from this location and be ready to repeat the operation farther on the course. The next time the bait is put down we should have plenty of bees in not more than ten minutes, and if they are tardy about coming, providing we had a fair amount at the first location, we have either passed the tree, are nearly under it, or have gone far off the course.

CHAPTER VI.

FALL HUNTING.

The main sources of the honey supply are now over, and if the methods given in the preceding chapters are followed it is necessary for us to get out on the mountains or fields far distant from home apiaries and look for the few flowers that have escaped killing frosts. A few bunches of mountain goldenrod are found here and there scattered over the mountain-side. A white flower, growing on a stem about two feet in height, is also found in many locations. I am unable to give the botanical name of this latter flower, but every bee hunter who has had much experience has seen many bees on it when other flowers have ceased to exist or have been rendered useless by frosts, as a source of honey.

If but a few of these flowers are found growing together and a few bees are seen on them, sprinkle freely with bait before described, and in a short time you will find ten bees to where there was one at first. Now if you start them from goldenrod, scent of almost anything used in bee hunting will serve to draw them on the course; but essence of goldenrod is far superior at this season of the year. As I have before stated, a scent should be used to conform as nearly as possible to the scent of the flower the bee is working on at any particular time. It would be a superfluity to explain any farther, as the same tactics must be followed as described earlier in this work.

CHAPTER VII.

THE LATEST IMPROVED METHOD OF BURNING.

We now come to the time of the year when all flowers, by the laws of nature, cease to bloom. Indian summer is here with its nice balmy days. Just right — not too warm not yet too cool. The very time when even those of us who are getting up in years begin to feel young again. How sad it would be to the one who loves nature and her ways to be obliged to lay aside all thought of sport until nature unfurled her robes again! Some of the happiest moments of my life have come during this part of the year, and I hope to be able to convince my readers that we should always say "welcome" to the aged year. Well do I remember when I used to go along with the old hunter in search of the bee. A fire would be made, some large flat stones heated and carried to a convenient place, then bee comb moistened with water, placed on them and soon bees would be seen darting through the air. Some might settle on the bait, but if not enough to satisfy the hunter, another hot stone was brought, and the process repeated until there were enough bees working on the bait to give a strong course. Then taking another hot stone and going a long ways on the course we would proceed to burn again. Perhaps the stone had cooled off by this time and the bee failed to come quickly or in sufficient numbers. Then we had to either go back, replenish the fire, heat more stones, or build another fire at the new location. Carrying the hot stones from place to place was the work generally assigned to me. Sometimes stones of a slaty nature would be heated and when becoming quite hot would burst with a loud report and fly in all directions. At that time I would just about as soon approach a loaded cannon. After twisting a stick around the stone it was carried at arm's length to the new location and with sweat streaming down my face I was glad when the time came to lay it down. This was undoubtedly laborious, but the excitement connected with the sport was at such a pitch that the thought of labor being in any way connected with bee hunting never entered my mind.

But as time wore on I got to thinking that there might be other plans much easier and quicker than the one described, and I feel sure that those who love the sport will agree that the plan laid before the readers is in every way superior to the old method

First get a small tin pail, holding about a half gallon. Cut out, from the bottom upwards, a hole four or five inches up and down and two inches wide. Have a pan made so that it will fit down inside the pail just deep enough to come down to upper edge of the hole cut out of pail. There should be a rim on top part of the pan to prevent it working lower down than the hole in the pail. Now get a miner's lamp, which will not cost more than from fifteen to twenty-five cents. Coal oil can be used but lard oil is much better, and better than either of these is alcohol. A small lamp suitable for burning this can be purchased at a small cost.

Now you are ready to start out. Take some refuse honey and your bottle of bait, get far out on the mountains, so there will be little danger of drawing bees from apiaries that may be situated in the valleys. When a suitable place is found, clear of underbrush and no large trees to bother the bees when starting for home, set pail down, put some of the honey in the upper part of the pail (or pan), strike a match, touch it to the wick of the lamp. The spout of the lamp should come within about two inches of the bottom of the pan. The honey begins to boil immediately and sends its scent out over the mountains. A few drops of the oil of anisc and bergamont mixed can be dropped into the pan, and a bunch of bushes held over the fumes until it is scented. This is then laid on the top of a bush or stump close by and sprinkled with bait. By this time bees may be heard darting through the air or seen hunting slowly through the bushes in search of something to eat. It is a very good plan to blow the lamp out when the first bees are flying around. The scent is strong all around and when the lamp is blown out the scent soon dies out except near the bait and the bees find the bait much sooner than if the lamp was kept burning. There may be plenty of bees to start with from the first burning and if not, all we have to do is to light the lamp again.

If you have your course and are about to start, it only requires a second of time to pick up the burning apparatus and the bunch of bushes and start on the course. But for fear you may be only a beginner and make a mistake which might discourage you, I want to have a little talk with you before starting from the first location.

In reading articles relating to bee hunting, some of the writers tell how, after loading up, the bees would circle round and round before starting on the homeward journey. I believe I have seen a few bees make a complete circle. I have seen hundreds of thousands that did not. As a rule when a bee raises from the bait it will act as though it intends to circle, but watch closely and you find before coming around to the place of starting it will quickly turn in the opposite direction, repeating this several times — always widening out. It will seem to fall far back with a downward motion, then gather up and come slowly back, often passing to the opposite side of the bait and making a sudden motion, is lost to sight. This fact might make you think the bee really went in this direction. I want to stake my reputation as a bee hunter of years of experience, that when a bee is seen to make these half circles on one side of the bait and seem to fall off in any direction, bearing down toward the earth, that this is the general direction in which the tree stands, and if I can see a bee make a few of these half circles (though it may be the first one on the bait), it settles the matter in my mind as to the general direction of the tree. But even if our minds are made up in regard to this line of flight, it is wise to take more time and watch closely, for there is no good reason why we should not get two or possibly more courses from this first location. Then go on the strongest course until we find the tree and then come back and start on the others.

In going on the course don't fail to look well at every tree, for sometimes they are found in very small trees when there are lots of large ones standing all around.

I will give my experience in finding a bee that has taught me to look at every thing on the course, not even discarded stumps, logs and bushes, for I have found bees in the two former and hanging on the latter. In early November I had a strong course from bait. They flew directly up on the side of the mountain. The course flew over a large barren thicket and after looking at the timber on the lower edge of the barrens, the bait was moved across the thicket. There were a few chestnut trees standing between the upper edge and the place I selected to bait them again. Soon they came and flew back down. I was sure they must be in one of the trees mentioned, for there was nothing growing in the thicket large enough for a bee to go in. After looking at the few trees spoken of and not finding them, I went back down to the lower edge and could see them fly nearly half way across the thicket. I was puzzled, and proceeded to look at the few logs that were laying down and still failed to locate them. My next move was to hang my burning bucket on a limb and burn. In no time there were bees by the quart on the bait, flying in all directions. Singling out some of the steady flying ones, they seemed to fly a short distance and drop into the brush. On investigating, I found them hanging on a little bush, working away as though they had the best place in the world to store their honey. They had evidently been there for a long time as they had several good sized combs fastened to the bush. I knew they were bound to perish, for cold weather was coming on, so I told a friend where to find it, and gave it to him with the understanding that he was to hive it, putting the combs and brood in the hive.

The above is mentioned to prove that bees are sometimes found in places out of the ordinary, and in closing this part of my work I want to impress you with the fact that it always pays to go slow and look well while on the course.

NOTE—If not convenient and a vessel of the kind described (for burning) cannot be had, any small tin pail will do without cutting out the hole for lamp. A couple of stones laid on the ground a few inches apart will make a place for the lamp and the bucket placed over it on the stones, although the first mentioned will be found more convenient.

CHAPTER VIII.

SOME FACTS ABOUT LINE OF FLIGHT.

You have all heard the term "bee line" used, and naturally infer that it means a straight line. This was what I believed it to be in my earlier days, but from numerous observations I am led to believe that the terms "bee line" and "straight line" are in some cases incompatible. If the line of flight is over ground unbroken by hills and hollows, a bee will fly as straight home after loading up as anything having wings can. But in following a course through a wooded country, along the side of hills or mountains containing ridges and deep hollows, the line of flight deviates far from a straight line.

To illustrate and prove the above assertion, I will here give an incident in connection with bee hunting that occurred not many years ago, and which goes to prove that bees do not always fly in a perfectly straight line. East of my home about one mile there is a mountain extending north and south. Along the foot of this mountain, a stream, known as Sideling Hill creek, runs the entire length of the valley. The mountain extending up from this creek is made up of ridges and hollows. A friend of mine, one day in July, found bees watering along the creek and nearly east of my home. The bees flew south with the creek along the foot of the mountain. After trying to find them, (consuming two days' time in the attempt), he came for me to help him out, telling me that he had looked at every tree near the course for a distance of a mile. It was a very finely marked Italian bee, and being anxious to find and hive it, offered to pay me for my time whether we found the bee or not. I asked him if he had baited them at the water. He said he had tried but not a bee could be induced to take bait. My time being limited just then, I told him I would get them to bait for him and after this he certainly could find it himself. "Oh, yes, that's all I ask," he replied. Going with him, I used the method described in an early chapter entitled "Hunting the Bee from Water." In a short space of time I had lots of them loading up and flying south along the creek. About a half mile on the course an old clearing ran up some distance on a ridge, and the course seemed to go about midway through it. My instructions were to put the bait on this place, as it was clear of all bushes that might bother him from getting a direct course, and after giving all necessary instruction I went home and awaited results. The next evening he told me he had gone into the old field and, as the bees were a little slow in coming to the bait, he built a fire and proceeded to burn and got bees in abundance, still flying on the same course; then moving the bait much farther on the course to another old field, found that they continued on the same line of flight; and from this last location followed them in sight of a house, the owner having thirty stands of bees, thus convincing him that the bees all had come from this apiary.

But I was convinced he had overlooked the bees started with, for these reasons: This apiary was two miles from where the bees watered; the same stream flowed near by the apiary — there were many springs near and water in abundance all along the course. Then the clearing first mentioned had lots of sumac growing in it; many bees from the apiary were working on this and other flowers, and by burning, these bees were enticed to the bait in such numbers that the few that may have been on bait from the tree were not noticed by an inexperienced hunter. After telling him of my suspicions, he was the more anxious that I should go along with him again and see for myself that there was no wild bee on the course.

I was equally anxious to prove to him that there was. So the following morning found us in the old field where he had first placed the bait. Taking my bottle containing bait. I sprinkled some on a bunch of bushes left there the day previous. This was all that was required and the bees that had been having a feast at this location the day before soon found it out and eagerly settled down for another feast. It seemed that the whole apiary had swarmed out and come to the bait — hundreds were soon flying towards this apiary. Here my friend ventured to ask if I was not convinced that they went to the apiary. I had been watching very close and knew very well that the majority of the bees did go there, but I had also seen a few bees fly a short distance on the course and bear off to the left. I said nothing about this at the time, thinking it best to be positive before giving a final opinion. There was a deep hollow running up from the opposite side of the clearing and getting in a more favorable position I could see many bees bear off from the main course and go up to the hollow. Now I was ready to tell him he had been outwitted by the bees.

Calling him to me, I showed him the bees flying up the hollow. We then moved the bait about one hundred yards farther up and found that they still went on up. We left the bait and proceeded to look at the timber. Finally one hundred yards above this last place there was a large white pine standing on the left side of the hollow and not over ten feet from the ground they were pouring in, in a steady stream, pure golden Italians. Was he convinced this was the bee we had started with from the watering place? No, not at all. It was too far from the course. I told him we would cut it and take it home, and if bees still continued to water at the same location I would give in. The bee was cut next day and taken home and all watering ceased at that place. This was evidence enough for him and proved to him, as it must to every one, that under certain conditions bees will vary very much from a straight line of flight.

CHAPTER IX.

BAITS AND SCENTS.

In rambling through the woods and over the mountains I have seen bee hunters using bait with the oil of anise in it, or perhaps a bait containing several different scents. They did not seem to know, nor care, that bait containing these oils was injurious to bees; but the fact is well known that they are injurious — not to our neighbor's bees alone, but to the ones we are trying to find. Therefore, never combine baits with scents of any kind. The former is intended to furnish feed for the bee, and when loaded will always start for the home. The latter is used as a means of getting them to come to bait.

There are many different scents used for enticing the bee to bait. Some hunters prefer oil of anise, others use bergamont; then some combine these or other scents. But bear in mind that what should be used ought to conform as nearly as possible in scent to the main source of nectar at any particular season of the year.

In preparing these scents, take an ounce of the oil you may prefer, put it into a pint bottle and fill bottle one-fourth full of alcohol; let it stand a few days and then fill up with water. This would make sufficient scent to last any one for several years. A small vial can be filled and taken along — even an ounce vial will last several trips; or a few drops of the oil can be put into a bottle and water added, but as water will not cut the oil, it remains insoluble and when the bottle is turned in order that the mixture will run out, it often happens that our scent (after using a time or two) is no good, the oil having disappeared. But by cutting the scent with alcohol, the last drop will be just as strongly scented as the first.

I have used about all the different scents known to bee hunters and oil of anise was my standby for many years. I found bergamont to be good. Horse mint, goldenrod, and many other oils and scents were used at some particular time of the year, but the most powerful and lasting scent I ever used was oil of sweet clover. Having run out of the oil and not knowing where to get it without sending to some drug house, I bought a toilet preparation labled "essence of sweet clover," and found it filled the bill. A few drops were spilled on my sleeve and in going on a course this was all that was needed. If I stopped but a moment, my arm was covered with bees.

I don't advocate the use of the hunting-box for bee hunting. I tried them long ago and found the method slow and uncertain. In carrying my box from one location to another and releasing the imprisoned bees I would always see them circle around and light on a leaf and consume from five minutes to a half hour in cleaning themselves up and when they did depart, there was no assurance that they would come back. However, some hunters must meet with better success than I have had in hunting by the box method, and to those I would say, if bringing the bees to your box is what you want, just rub a few drops of the oil of sweet clover on the side of your box and that part of finding the bee is done.

It is hardly necessary to say more about baits. My views have been given in the earlier chapters on bee hunting. A few drops of pure honey is perhaps the best that can be used in starting the bees on bait, but as soon as several have loaded with the honey, sprinkle your bunch of bushes which you intend to carry on the course with a bait made by filling a bottle one-fourth full of pure granulated sugar, then a little honey and filling the bottle up with water. This will make the bait sweet enough and it will not become so sticky as if more sugar or honey were used.

CHAPTER X.

CUTTING THE TREE AND TRANSFERRING.

I hope those who read this book may find something in its pages that will be beneficial. In your excursions through the forests you are unconsciously getting the benefit of the greatest source in the world of physical perfection — God's pure air — and, at the same time there are no reasons why one with reasonable tact cannot be benefited financially.

When should a bee tree be cut and transferred to the hive? There is a difference of opinion in regard to the time of the year and also to the manner in which it should be done. I respect the opinions of those who have expressed themselves on the subject, but after trying nearly all the methods described I found nothing in them that came up to my ideal of a perfect plan of transferring the bee from the tree to the hive.

My first plan was to cut the tree and, if not too large, saw it off both above and below the bees, keep them in with smoke, and tack screen over the place of entrance. Then hire someone to help carry it home. It was set up on end and left to take care of itself and if a swarm would issue from it and we were successful in hiving it in the old box hive (the kind mostly in use in my boyhood days), we thought the last chapter of bee-keeping had been learned. Then, after the movable frame hive came into use the tree would be cut, the bees drove into a box, the honey taken from the tree and with a few pieces of brood all was taken home. The small bits of comb were tied in the central frames for the bees to cluster on and the bees shaken from the box in front of the hive. This plan was certainly superior to the first mentioned but had one serious drawback — the brood that was in the tree was left to perish.

After seeing the serious defects in the described methods, my next move was to take a hive with me on going to cut the tree. All comb containing brood was placed in the frames, the bees run into the hive, which was left at the tree for a week or more in order that the bees might have all the combs joined to the frames, and then brought home. This was another advance in the method of transferring, for the thousands of young bees about to emerge from their cells were saved, and the colony having its brood and strength undiminished should be able to fill at least one super of honey besides all stores needed for themselves. Taking it for granted that we cut the bee in the early part of the summer, one super would be a low estimate, but even this would pay all expenses connected with the cutting, buying a hive and fixtures, and as the bee is now in an ideal hive we can hopefully look forward to the next year when our profits are coming in.

There could be other plans given, some of them having virtue, but I will now lay a plan before the reader which if followed will prove more remunerative, and with less expense, than the former methods. To carry a hive and tools necessary to cut a bee tree will require the service of an assistant and when, after a week or so, we return to bring the bee home, more help is needed. A man is worthy of his hire and of course is paid. Carrying a hive over rough and uneven ground is hard work. So by the time we have the bee home and sum the matter up, the financial part of bee hunting don't impress us very strongly.

I have been in the habit of hunting bees during the fall months, but if I need a day's outing, no month from early spring, until late fall fails to find me on my tramps through the forest in search of a bee tree. No difference what time of the year I find my bee nor how many may be found in any particular season, they are always left stand over winter and cut the following spring, but not before May, for I want the bee to be strong in bee with abundance of brood. About this time of year I take a box eight inches square at the end and two feet in length. Over the one end some wire screen is nailed and a lid, the center being cut out and replaced with wire screen, serves as a covering for the other end.

With bucket, ax, and this box we will go to the tree, cut it, being careful to fell it as easy as possible. When it falls the bees should be smoked at once to prevent them rising in the air. For good reasons I prefer to cut the tree about nine or ten o'clock in the forenoon. After blowing a little smoke in at the entrance, proceed to chop a hole in the tree low down on the side, then another hole farther up or down the tree, depending on whether the bee works up or down from the place of entrance. After this is done, split the piece out, blow more smoke on the bees and take the combs out. Brush the bees off, lay them on the log some distance from the bees, place the forcing box over the main body of the bees and by brushing and smoking drive them into it. The box should be in an elevated position, say forty-five degrees or more, as bees will go on the upper end much more readily when the box is in this position. Be sure the queen is in, which can generally be determined by the manner in which the bees enter the box. If they are inclined to run back out after being forced in, it is a pretty sure sign the queen is not with them. When you are sure the queen is with them, and there is a sufficient number of bees with her, lift the box gently off, turn it upside down and place the lid on and fasten with a couple of tacks taken along. Now place the brood combs back in the tree. First a comb then a couple of small sticks crosswise to form a bee space. Continue this until all the combs are back in the tree, and as the top part of the log was not split off, the piece split from the side can be fit in, bark and flat stones can be used to form a covering that will keep the rain from getting in. By cutting the tree at this time of day thousands of bees are out in search of nectar and when they come home and find their home gone, will fly around in the air until becoming exhausted, and will then settle on the leaves and bushes in bunches and knots by the hundreds. If there was any nice white honey we have it in the bucket and picking up the box start on the homeward journey. Presuming we have a movable frame hive at home with an inch of starter in the frames or, what would be better, a hive filled with comb from the year previous, we place the hive on its permanent stand and take the lid from the box and shake the bees down at the entrance. For fear the queen has been left in the tree it would be well to have an entrance guard placed on the hive, as this would exclude the queen and as soon as the queen is seen the guard can be removed. In a short time we can tell whether they take kindly to their new home. The queen is a laying one and some pollen should be taken in the following day. I always made sure I had the queen and never had a bee so treated to swarm cut after being hived.

Now what about the bee in the tree? When we left it there were thousands flying around and settling on the leaves and bushes, other thousands in all stages of development in the combs. The ones that are hanging on the bushes begin to make further investigation and finding their brood soon cover it and with the bees hatching out every hour soon make the colony almost as populous as it was before the tree was cut. In taking the combs out we may have seen some queen cells started. If so, so much the better. If not, there certainly were eggs in some of the combs and in sixteen days at the most they can rear a queen from these eggs. When this time has elapsed, take your box and smoker. Take the combs out as before; drive the bees into the box, and as the brood is nearly all hatched out by this time you will have nearly as many bees as you got the first time. These are brought home and treated as the first swarm and the combs can be placed in the log again for the few remaining bees that may have been left, to cluster on and these can be brought home later and joined to the second swarm. By this method you get two strong colonies from one tree. There is no help needed; no neavy lifting and carrying of hives to and from the tree. By following this plan you can soon have quite an apiary and be on your way to enjoy the profits as well as the pleasures of bee hunting. This plan is original with me and I believe it to be the very best plan given so far, and I expect to follow it until someone gives us something superior.

The profits of bee hunting will depend on the ability of the man to manipulate the bees after taking them from the tree. You must agree with me that in cutting the tree, there is nearly always some of the combs containing honey broken up and covered with dirt, and this honey can never be classed as salable. Therefore, if we hunt bees merely for what honey may be in the tree and leave the bees to perish from starvation and cold, it were far better, from a moral and financial point of view, to let the tree stand.

CHAPTER XI.

CUSTOMS AND OWNERSHIP OF WILD BEES.

There are customs in vogue among sportsmen that have been handed down from generation to generation, that have almost become laws. Indeed, we have heard it said that custom becomes law.

A hunter may wound a deer, follow it for a distance and find that another hunter has shot and killed it. The question might arise as to whom the deer belonged. A bee hunter may find a bee tree and mark it and some other hunter might find it afterwards and cut it. The same question might arise as to whom it legally belonged. If sportsmen were to settle the disputes they would refer back to custom and say the deer belonged to the one first wounding it, providing the wound was of such nature that the one first wounding it would have been pretty sure of getting it, by following on, and they would also decide that the bee belonged to the one who first found and marked it.

A custom that may seem to be founded on justice is pretty apt to be followed by laws that may coincide with the custom. But we must remember there are statute laws relating to the ownership of wild animals and bees, and though we all band together as sportsmen, we cannot abrogate nor set aside these laws already formed.

In my boyhood days, when I would find a bee, I was very slow to tell any one just where it was for fear they might cut it. Was this true sportsmanship? I think not. Some other bee hunter might hunt for that bee a day or more and finding it would have reason to say that I had deceived him and he could hardly be blamed if he cut it. I have been used just this very way more than once, and felt like retaliating by cutting a bee that was found prior by another party. But am glad to say that I never did. Since I became more mature in years I have had more confidence in my fellow sportsmen and now after finding a bee tree the first time I see any one who is likely to look for the bee, he is told its exact location, thus probably saving him much valuable time in not looking for a bee that is found.

As a fitting close to this work it might be well to quote the statute laws relating to the ownership of wild bees.

"Bees while unreclaimed, are by nature wild animals. Those which take up their abode in a tree belong to the owner of the soil, if unreclaimed, but if reclaimed and identified, they belong to the former owner. If a swarm leave a hive they belong to the owner as long as they are in sight and are easily taken; otherwise they become the property of the first occupant. Merely finding a bee on the land of another and marking the tree does not vest the property of the bees in the finder. *They do not become private property until they are in a hive.*"

This is a statute law. But true sportsmen do not think of going to law for adjustment of these matters, but rather depend on that fraternal spirit by which all questions relating to ownership are settled amicably.

CHAPTER XII.

SOME OF OUR BENEFACTORS AND THEIR INVENTIONS.

Bee keeping as a source of revenue dates far back in ancient history. With the advent of the movable frame hive and the increased demand for honey all over the world as a source of food supply, it received a new impetus and there are many bee keepers in this and other countries who are not only making an honest living in the pursuit, but have become wealthy as well.

Over half a century ago, Rev. L. L. Langstroth invented the movable frame hive and became the benefactor of the bee-keeping fraternity. Prior to this time there was no way of telling the condition of a bee except what could be learned from an external diagnosis. If from their actions we were led to believe the colony was diseased, or that the bee moth was holding sway, there was no way by which we could remedy the evil. But this invention gives us access at all times to the brood chamber and we are able to see just what is wrong and apply the proper remedy. Perhaps it is fair to add that all bee keepers do not agree that the movable frame was invented by Father Langstroth. This honor is conceded by many to belong to Huber or Dzierzon, German bee keepers. Be this as it may, the movable frame hive of today, used throughout America and many foreign countries, is the product of the inventive genius of this great benefactor of the bee-keeping fraternity.

The invention of many accessories since the death of Father Langstroth, many years ago, would almost make us believe that there is nothing further to be desired, that perfection has been reached. But well we know that perfection cannot be reached on this earth, and so we will look forward, knowing as time goes on that other great minds will add to the store of knowledge now possessed by the bee keeper, and bee keeping of the future will be as far in advance of the present as the present is of the past.

With the help of appliances and the instruction given by able writers in many magazines and bee papers anyone with a fair amount of ability should be able to make a success at this vocation. There are many men who, while they have proved to be benefactors to us, have at the same time become wealthy. There are many instances of this, but I will mention The A. I. Root Co., of Medina, O. A. I. Root, the senior member of this firm, was an apiarist of note while I was still a little boy. After a while he began the manufacture of hives and appliances. He invented the pound section box, the extractor and many other accessories that could not be dispensed with at the present day. Many of his inventions were never patented, thus saving that cost to those whom he wished to befriend, and by honest dealing, selling the best of everything needed by the apiarist at the lowest possible cost consistent with superior workmanship, he has today, the most extensive manufacturing establishment in America, and possibly the world. In connection, the firm publishes, "Gleanings in Bee Culture," a monthly magazine, devoted to the interest of bee keeping. The ablest writers, men who have made this their life work, contribute regularly and give us advice which, if followed will lead to success.

Therefore, when the bee history is completed, and the names of many who have been our benefactors are recorded, the names of L. L. Langstroth and A. I. Root will shine with lustre.

CHAPTER XIII.

BEEKEEPING FOR PROFIT.

It is not generally known that beekeeping is quite an industry in the United States and that this country maintains a lead over all other lands both as to the quantity and quality of the honey it produces. This is the case, however, and America is recognized by other countries as the honey-land par excellence, where beekeepers turn out honey by the carload and this is so, for California, in one lone year, produced 800 carloads, and of this 500 were shipped out of the state. Texas is also a heavy producer and year in and year out will actually outrank California.

Although produced in such vast quantities it must not be inferred that quality is neglected; on the contrary we cannot be excelled when merit is considered. Our apiarists are scientific to a very high degree and possibly no branch of American farming has been worked up to so great a pitch of excellence, only dairying and horsebreeding can be compared with it. but American apiculturists lead the world, whereas, our horsemen or dairy-men do not.

This proud position is owing to the splendid discoveries and inventions of the Rev. L. L. Langstroth of Oxford, Ohio, who has been dead for some years, but whose spirit still lives. Previous to his time beekeeping was only an amusement or pastime, or more accurately speaking, a hobby.

Now, the industry is founded on a sound scientific basis and bids fair to grow at a lively rate in the years that are to come. At present, the amount of money invested in bees and bee appliances is not less than one hundred million dollars. The annual income from this source cannot be much less than $20,000,000, and in a good year all over the country, it would approximate $50,-000,000 though it is very seldom that there is a good season for bees all over this vast country. Beekeeping is a branch of agriculture and like other pursuits belonging to that science there are fat years and lean years. It is not an uncommon event for a beekeeper to clean up a sum of money for his crop which will more than equal the value of his bees and all the appliances he uses. Other years may be total failures, but year in and year out no industry pays larger returns on the labor and money expended. The wise beekeeper is not deterred by a bad season but simply bides his chance. He knows that in course of time the bees will make good all losses and give in addition a handsome profit to the owner for his kind attention and thoughtful consideration.

There are still many opportunities for beekeepers in this country. This is particularly true of West Virginia, Tennessee and Kentucky, where the conditions for beekeeping are almost ideal and where, as a usual thing, the market for honey is good. All through the South there are openings for beekeepers and it will be a long time yet before all openings are filled. Southwest Texas is a sort of beekeeper's paradise and only a part of it has been occupied as yet. Arkansas is a particularly good state for bees, but it has only been partially developed by up-to-date beekeepers. Parts of Pennsylvania are open to good beekeepers and so are portions of Michigan, one of the leading states of the Union. Ontario and Quebec are excellent for bees — none better. Nearly all the western states are good for bees and some of them rank high as honey producers. This is true of Colorado and Utah. Idaho, Montana, Nevada, Wyoming, New Mexico, Arizona, Washington and Oregon offer excellent openings for first-class beekeepers. In the West, beekeepers, usually select an irrigated region where alfalfa and sweet clover are common, so that during the long dry summers the bees are kept busy storing honey of a very high quality.

Successful beekeepers are found in every state, and it would be hazardous for anyone to say just what state is best for bees. Ohio, Indiana and Illinois produce large quantities of fine honey, but this is nearly all consumed within their own borders at fair prices so that beekeepers do fairly well.

What hinders beekeeping more than any other fault is the neglect of the beekeepers in not providing adequate shelter for the bees during cold weather, and also from the heat of summer. In the Northern and Central states good protection must be provided against zero weather. Our bees originally came from the tropics, and for that reason they require ample protection. The ordinary hives must have an outer case placed around them and then leaves, straw or sawdust well packed around them. Fixed in this way they will withstand the rigors of an arctic winter. Lack of adequate winter protection is the weakest point in American bee culture, and yet is easily provided. This accounts for the saying of many who have tried it, "Beekeeping doesn't pay." Perhaps at no time is protection more necessary than in early spring when the hives are full of young and tender brood. The hives may also be covered with layers of thick paper or asbestos board. A small hole will allow all of the fresh air necessary for bees in a state of sleep. These points are first mentioned because neglect of them accounts for most of the failures we often hear of.

No success can be anticipated unless one uses the best hives made on the Langstroth principle. We have no space here in which to give a complete account of the hives now made on that plan. The better way would be for anyone interested to write for a sample of "Gleanings in Bee Culture" Medina, Ohio, or to American Bee Journal, Hamilton, Illinois, so as to get in touch with the publishers, who issue books adapted to the wants of beginners. These magazines also issue supply catalogues and in other ways are quite helpful. Splendid books can be purchased at a low price giving complete information with regard to the bee industry. Many persons have learned the whole art of beekeeping by a careful study of a good book on bee culture supplemented of course by observation.

Nothing very important, however, can be learned about bees unless one possesses a colony of bees in a movable comb hive. In fact it is useless to attempt to obtain a knowledge of bees without a hive to work with. I, therefore, earnestly recommend any beginner to obtain a colony at the earliest opportunity. Very often an ordinary box hive can be secured for a "song." This will do to begin with. Next send for two complete standard Langstroth hives, a smoker, a veil and a bee book; also a swarm-catcher.

If the box hive is of a medium size it will probably cast two swarms in spring about fruit-bloom time or a little later. When the swarms emerge they may be quickly taken down by means of the swarm-catcher, if they happen to lodge in a branch of a tree, as they usually do. If the hives are in readiness it is no great feat to safely place the swarms in their new homes and all will go well. The parent colony may be disposed of in a week or ten days (not later) after the second swarm issues, by drumming the bees out of the box into the hive which holds the second swarm. This is done by giving them smoke from the smoker and then battering on the hive with a stick, which so alarms the inmates that they rush over the side of the upturned hive into the new one. What is left is simply a lot of dirty combs fit only for the melting pot. This is probably, the neatest, cleanest and cheapest method of making a start in beekeeping. It is well within the ability of most men and the cost is comparatively small. If the bees are native blacks, later on they may be changed to Italians simply by purchasing young pure bred queens for about a dollar each. The old queens are killed and new ones introduced in a cage till the bees make her acquaintance, when she is automatically released. In two months' time very few of the original bees will be found, all having died from hard work and old age, and their places taken by rich golden yellow Italian bees. It may be well to add this caution, "Do not experiment with any other race of bees."